I0883462

20-MINUTE (OR LESS)

DIGITAL STORYTELLING HACKS

SHEELA PREUITT

Lerner Publications ◆ Minneapolis

To Ana & Samru, with all my love.

Online Safety

Children should always ask permission before going online, especially when using a new website. Stay safe online by protecting your identity. Never share personal information such as your full name, address, or email address online. Think carefully about posting personal photos of yourself, and don't post pictures of friends or family without permission.

Copyright © 2021 by Lerner Publishing Group, Inc.

All rights reserved. International copyright secured. No part of this book may be reproduced, stored in a retrieval system, or transmitted in any form or by any means—electronic, mechanical, photocopying, recording, or otherwise—without the prior written permission of Lerner Publishing Group, Inc., except for the inclusion of brief quotations in an acknowledged review.

Lerner Publications Company
An imprint of Lerner Publishing Group, Inc.
241 First Avenue North
Minneapolis, MN 55401 USA

For reading levels and more information, look up this title at www.lernerbooks.com

Main body text set in Conduit ITC Std Regular 13/17
Typeface provided by ITC

Library of Congress Cataloging-in-Publication Data
Names: Preuitt, Sheela, author.
Title: 20-minute (or less) digital storytelling hacks / Sheela Preuitt.
Other titles: Twenty minute (or less) digital storytelling hacks
Description: Minneapolis : Lerner Publications, 2020. | Series: Vidcode coding hacks | Includes bibliographical references and index. | Audience: Ages 8–12. | Audience: Grades 4–6. | Summary: "By completing these projects, readers will gain a solid baseline in coding. Build a slideshow, design a flag, create headlines that scroll across the bottom of newscasts, and more!"— Provided by publisher.
Identifiers: LCCN 2019052412 (print) | LCCN 2019052413 (ebook) | ISBN 9781541597228 (library binding) | ISBN 9781728401423 (ebook)
Subjects: LCSH: Digital storytelling—Juvenile literature. | Interactive multimedia—Juvenile literature. | Computer programming—Juvenile literature. | Storytelling—Juvenile literature.
Classification: LCC QA76.76.I59 P74 2020 (print) | LCC QA76.76.I59 (ebook) | DDC 808.5/430285—dc23

LC record available at https://lccn.loc.gov/2019052412
LC ebook record available at https://lccn.loc.gov/2019052413

Manufactured in the United States of America
1 – CG – 7/15/20

TABLE OF CONTENTS

CODE YOUR STORY

Don't you love a good story? You can make a digital story in 20 minutes or less by **coding**! Digital storytelling combines art and science to make a **narrative** about any topic. Using the Vidcode platform, you can make digital stories with just a few lines of code. You'll be **programming** before you know it!

Coding takes practice. As you learn, you might give the computer a wrong instruction and end up with something you don't like. When this happens, don't worry! Stop, think, and change your code to get the result you want.

Ready to show off your creativity?

To complete each project in this book, you can work online at the Vidcode Lerner sandbox. You will be able to save and customize your personal projects.

vidcode

GO HERE TO GET STARTED!

qrs.lernerbooks.com/Storytelling

SLIDESHOW

JavaScript is a **programming language**. Use JavaScript to tell a computer what to do and become a programmer!

PROJECT TIME: 15 MINUTES

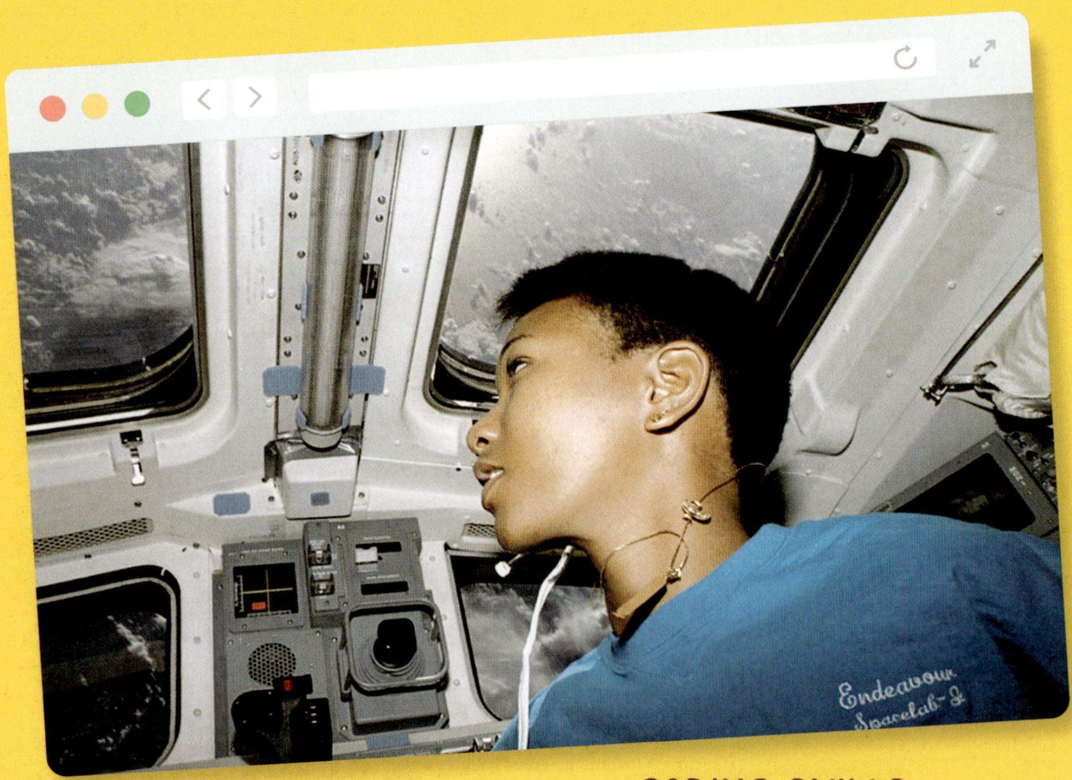

CODING SKILLS

- arrays

PROJECTS:
qrs.lernerbooks.com/Storytelling

Click the Back and Next buttons in the Instructions panel at any time to unlock effects and access useful tips and tricks.

1. From the dropdown menu in the **Frames** tab, select "People."

2. Click on a **thumbnail** image from the Frames tab. The movie file will appear on the **canvas** as two lines of code in the **code editor**.

3. Click on more thumbnail images. The names of the images will appear listed in brackets in the code. This list is called an **array**.

4. In the code editor, type the line:
`movie.interval = 500;`
In this code, `movie` is an **object** and `interval` is a **property**. Properties add information about how an object looks or behaves.

5. The interval **value** in the new line of code is `500`. Make this value larger or smaller to make the animation run faster or slower.

6. Mix up your slideshow by changing the list order inside the brackets or adding **filters** from the Effects tab.

1

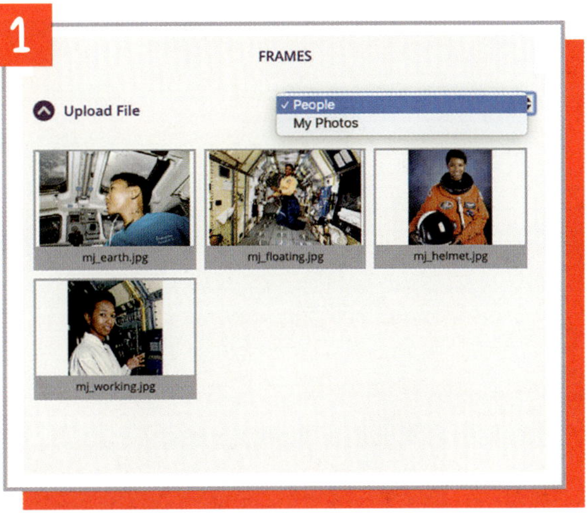

4

```
movie = stopmotion();

movie.frames = ["mj_earth.jpg",
                "mj_floating.jpg"];

movie.interval = 500;
```

DIGITAL QUILT

PROJECT TIME:
10 MINUTES

Create a unique filter using the **pixelate** function. Just like a quilt is made of different fabrics, your digital quilt will be made up of blocks of colored pixels!

CODING SKILLS

- functions
- values

PROJECTS:

qrs.lernerbooks.com/Storytelling

1. Click on a video in Backgrounds.
2. From Effects, drag the pixelate **function** and drop it into the code editor. Your video now displays as square blocks of color.
3. Drag and drop other functions from Effects. Watch what happens to the video as you stack multiple functions.
4. Delete the code and repeat steps 1 through 3. Change the order in which you add functions until you create something you love!

1

3

```
movie = video();
movie.source = "lights.mp4";

pixelate(50);

color_invert();
vignette(50);
noise(50);
```

4

```
movie = video();
movie.source = "lights.mp4";

pixelate(50);

color_invert();
vignette(20);
noise(50);
```

CODING TIP!

Adjust the value of a function by clicking on its line of code and changing the number. Watch how changing these values changes your digital quilt.

TRICOLOR FLAG

Create a digital flag with different colored stripes.

PROJECT TIME: 15 MINUTES

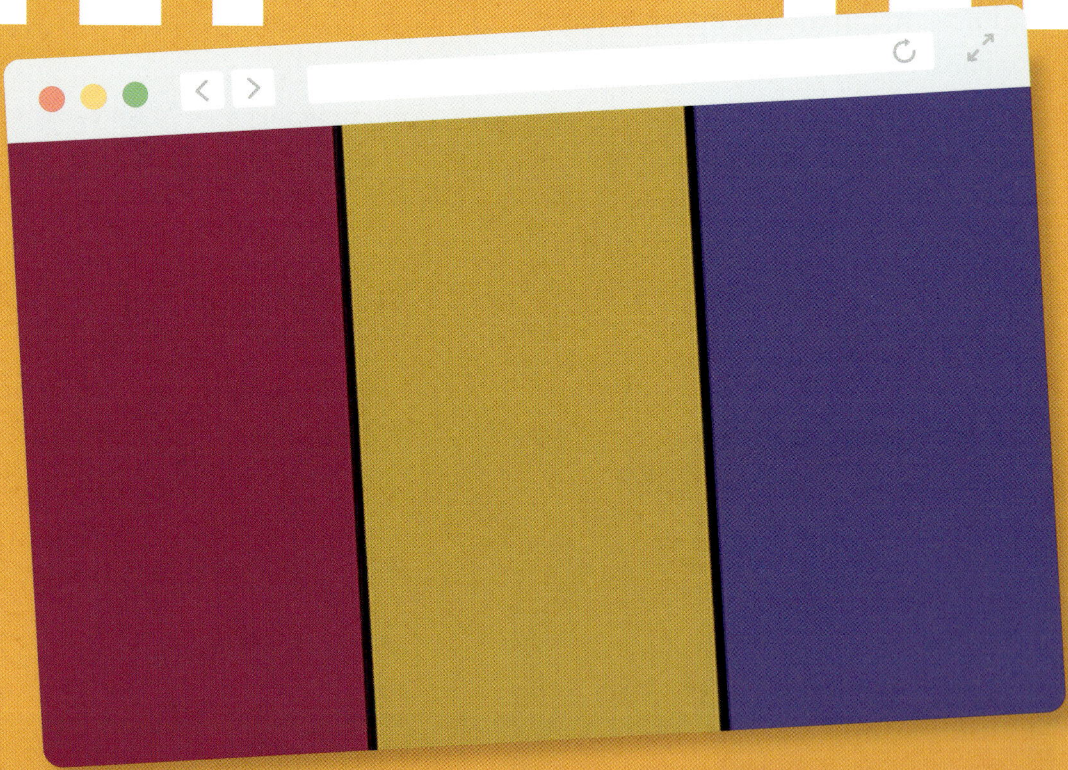

CODING SKILLS

- variables
- properties
- values

PROJECTS:
qrs.lernerbooks.com/Storytelling

1. Drag a rectangle block from the Effects tab into your code editor.

2. The rectangle will appear on the canvas and as one line of code in the code editor:
```
var my_rect = rect(0, 0,
150, 200);
```
In this code, `var` stands for **variable** and `my_rect` is the variable name.

3. The function `rect(0, 0, 150, 200);` has four **arguments** in parentheses. The first and second refer to the 'x' and 'y' position of the top left corner of the rectangle. The third argument is the rectangle's width, and the fourth argument is the rectangle's height. Click in the code and change these values. Watch how the position and size of the rectangle change too!

4. Change the color of your rectangle by adding another argument to the parentheses:
```
var my_rect = rect(0, 0,
150, 200, "red");
```

5. Type in a new line of code to set the **opacity** property. The opacity value must be between 0 and 1:

```
my_rect.opacity = 0.5;
```

6. Change the width value of the rectangle to `movie.width/3` and the height value to `movie.height`. This makes your rectangle cover the left third of the canvas.

7. Drag another rectangle block from the Effects tab into the code editor. Click in the code to change the rectangle's variable name to `my_rect2`.

8. Change the argument values in parentheses to create the middle third of your flag:

```
var my_rect2 = rect(movie.
width/3, 0, movie.width/3,
movie.height);
```

9. Change the color and opacity for `my_rect2`.

6

```
var my_rect = rect(0, 0, movie.width/3,
                         movie.height, "red");

my_rect.opacity = 0.5;
```

7

```
var my_rect = rect(0, 0, movie.width/3,
                         movie.height, "red");

my_rect.opacity = 0.5;

var my_rect2 = rect(0, 0, 150, 200);
```

8

9

10. Drag a third rectangle block from the Effects tab into the code editor. Change the rectangle's variable name to `my_rect3`.

11. Change the argument values in parentheses to create the final stripe of your flag:

```
var my_rect3 =
rect(2*movie.width/3,
0, movie.width/3, movie.
height);
```

12. Change the color and opacity for `my_rect3`.
You coded a tri-color flag!

10

```
var my_rect = rect(0, 0, movie.width/3,
                        movie.height, "red");

my_rect.opacity = 0.5;

var my_rect2 = rect(movie.width/3, 0,
   movie.width/3, movie.height, "yellow");

my_rect2.opacity = 0.8;

var my_rect3 = rect(0, 0, 150, 200);
```

12

CODING TIP!

The symbol * is the multiplication **operator**. It multiplies two numbers together, like the symbol x in math. The symbol / is the division operator. It divides a number by another number, like the symbol ÷ in math.

BEHIND THE CODE

It may not be obvious at first what your code is instructing. But every line of code has a purpose!

The full width of a canvas can be refered to as `movie.width`.

The full height of a canvas can be refered to as `movie.height`.

```
var my_rect1 = rect(0, 0, movie.width/4, movie.height, "#C3F2EC");
```

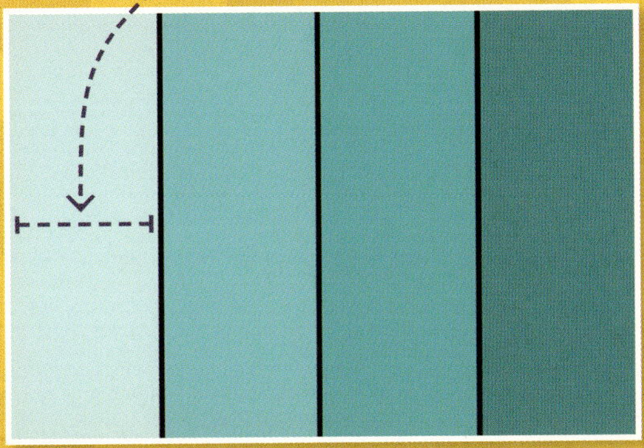

By using the division operator `/4` here, the width of the rectangle is set to be one-quarter the width of the canvas. If you were to use the code `movie.width/5`, the rectangle would be one-fifth the width of the canvas.

```
var my_rect2 = rect(movie.width/4, 0, movie.width/4, movie.height, "#5FDFDA");
```

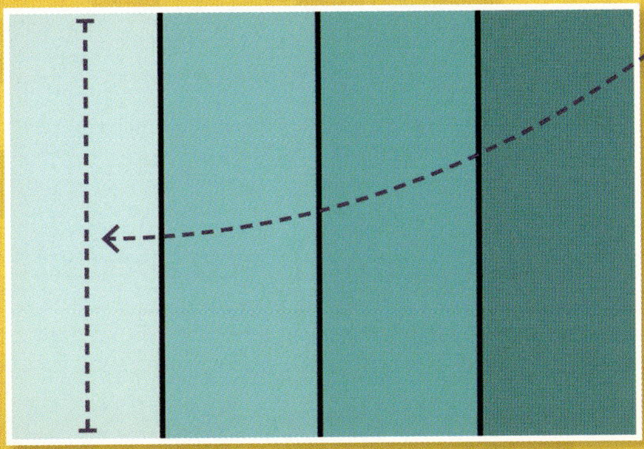

The code `movie.height` makes a shape stretch the whole height of the canvas.

MAKE NOISE

PROJECT TIME:
15 MINUTES

Learn how **loops** and a filter effect called noise affect a video!

CODING SKILLS

- operators
- loops
- variables

PROJECTS:
qrs.lernerbooks.com/Storytelling

1. Click to select a video from the Backgrounds tab.
2. Drag the noise block from the Effects tab into your code editor.
3. Change the noise value of `50` to any number between `1` and `100`. Start with a very low number at first, such as 5.
4. Type `var my_noise =` in front of the line of code from step 2 to make a variable. In this code, `var` stands for variable, `my_noise` is the variable name, and `noise` is the function the variable is holding.
5. Drag a repeat block into the code editor from the Effects tab.
6. Inside the curly braces, type `my_noise.amount++;`
7. As the video repeats, watch how the noise increases. To make this happen faster or slower, change the number 5 at the bottom of the repeat block.

4

```
movie = video();
movie.source =
  "monarch-butterfly.mp4";

var my_noise = noise(5);

```

6

```
movie = video();
movie.source =
  "monarch-butterfly.mp4";

var my_noise = noise(5);

repeat(function() {
    //write code below
my_noise.amount++;
}, 5);
```

HACK IT!

This project uses the ++ operator. It signals for a value to increase by 1 each time a line of code runs.

CODE THE NEWS

Tell the world your news by coding scrolling news bulletins, just like on TV!

PROJECT TIME: 15 MINUTES

CODING SKILLS

- loops
- operators

PROJECTS:

qrs.lernerbooks.com/Storytelling

1. Click on a news video from the Background tab. The video's code will appear as two lines of text in the code editor.

2. Drag a text block from the Effects tab into the code editor. The text block will appear in the code editor as this line of code:

```
var my_text1 =
text("hello");
```

Here, `var` stands for variable and `my_text1` is the variable name.

3. Change the variable name to `headline` within the code.

4. Delete the word "hello" and write a funny news bulletin in its place. Do not delete the quotation marks.

5. Change the color and size of your text by adding and adjusting properties. Add the code below to experiment:

```
headline.color = "yellow";
headline.size = 35;
```

2

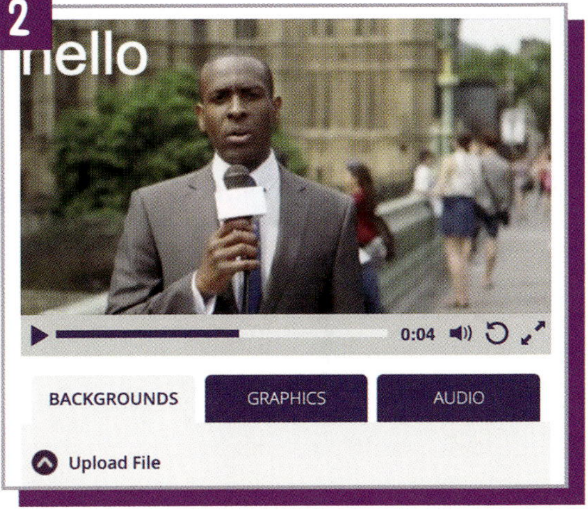

3

```
movie = video();
movie.source = "anchor_2.mov";

var headline = text("hello");
```

4

```
movie = video();
movie.source = "anchor_2.mov";

var headline =
  text("Breaking News! Free Cake!");
```

6. Change the 'x' and 'y' properties to position your text along the bottom of the screen. Add these lines of code to the code editor:

```
headline.x = 10;
headline.y = 212;
```

7. Drag a rectangle block from the Effects tab into the code editor. The rectangle will appear in the code editor as one line of code:

```
var my_rect = rect(0, 0,
150, 200);
```

The arguments in parentheses refer to the 'x' position on the canvas, the 'y' position on the canvas, the width of the rectangle, and the height of the rectangle.

8. Change the rectangle's width value from `150` to `movie.width` to make your rectangle fill the width of the canvas.

6

```
movie = video();
movie.source = "anchor_2.mov";

var headline =
  text("Breaking News! Free Cake!");

headline.color = "yellow";
headline.size = 35;

headline.x = 10;
headline.y = 212;
```

7

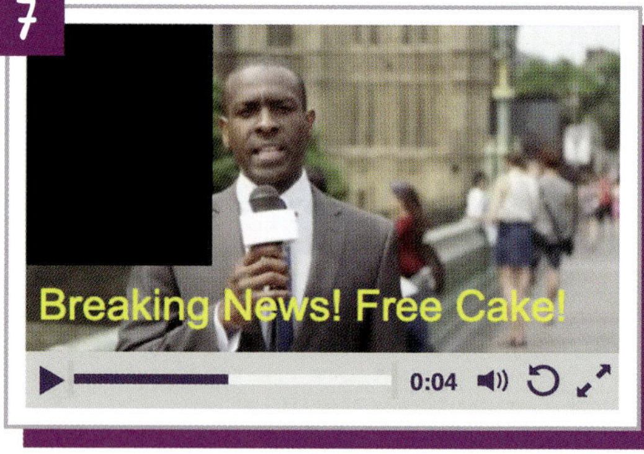

8

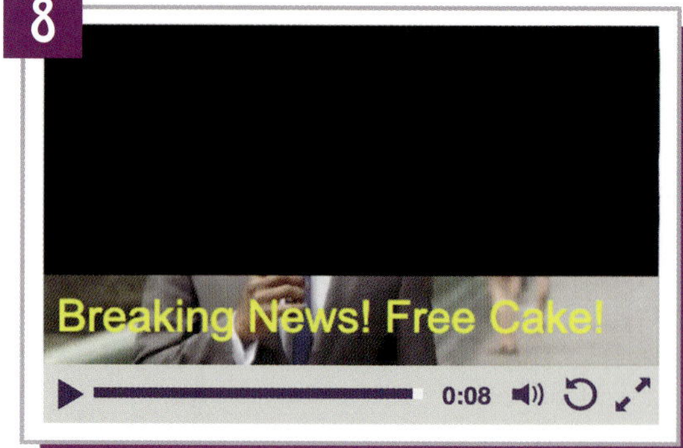

9. Click in the code to change the name of the variable from `my_rect` to `headlineBox`. Then, cut and paste this line of code so that it appears above the line of code that begins `var headline`.

10. Change the other values within the parentheses to position the box at the bottom of the screen behind your news bulletin.

11. Change the color and border color of the rectangle by adding fifth and sixth values to the function parentheses:

```
var headlineBox = rect
(2, 200, movie.width, 71,
"black", "yellow")
```

12. Drag a repeat block from the Effects tab into the code editor.

13. To make your bulletin scroll from right to left, type the line of code inside the repeat block beneath the line

```
//write code below:
headline.x -= 20;
```

This will make the 'x' position of your text repeatedly move to the left by 20 pixels!

9

```
movie = video();
movie.source = "anchor_2.mov";

var headlineBox =
  rect(0, 0, movie.width, 200);
var headline =
  text("Breaking News! Free Cake!");

headline.color = "yellow";
headline.size = 35;

headline.x = 10;
headline.y = 212;

```

10

13

```
movie = video();
movie.source = "anchor_2.mov";

var headlineBox =
  rect(2, 200, movie.width, 71,
       "black", "yellow");
var headline =
  text("Breaking News! Free Cake!");

headline.color = "yellow";
headline.size = 35;

headline.x = 10;
headline.y = 212;

repeat(function() {
    //write code below
    headline.x -= 20;
}, 5);
```

14. Drag an if block from the Effects tab and place it inside the repeat block from steps 12 and 13.

14

```
repeat(function() {
    //write code below
    headline.x -= 20;
    if (true) {

    }
}, 5);
```

15. Replace the word "true" in the if block code with `headline.x < -500`.

16. Directly under the code from step 15, add the line of code :

```
headline.x = 500;
```

The full block of code will look like this:

```
repeat(function() {
 //write code below
headline.x -= 20;
if (headline.x < -500) {
headline.x = 500;
}
}, 5)
```

16

```
movie = video();
movie.source = "anchor_2.mov";

var headlineBox =
  rect(2, 200, movie.width, 71,
       "black", "yellow");
var headline =
  text("Breaking News! Free Cake!");

headline.color = "yellow";
headline.size = 35;

headline.x = 10;
headline.y = 212;

repeat(function() {
    //write code below
    headline.x -= 20;
    if (headline.x < -500) {
    headline.x = 500;
    }
}, 5);
```

17. The if block is looping your news bulletin by repeatedly moving the 'x' position of the text. The headline moves back to the right of the canvas each time it disappears offscreen.

17

CODING TIP!

Code is read from top to bottom. This order affects the layering of objects. For text to appear on top of a rectangle, its code must be written after the rectangle's code in the code editor. This is a **sequence**.

VIDEO DJ

PROJECT TIME: 20 MINUTES

Create a music video that changes tracks as you tap your keyboard!

Pressing a key on the keyboard or clicking a button on a web page are "events." JavaScript can be coded to respond to events in a specific way. In this project, when any key is pressed, JavaScript changes the audio file.

CODING SKILLS

- audio
- keyboard events

PROJECTS:
qrs.lernerbooks.com/Storytelling

1. Click on a video from the Backgrounds tab.

2. Click on any music file except "retro.mp3" from the Audio tab. The audio file will appear as one line of code in the code editor. The music will play automatically with your video.

3. On the same line as the audio file name, type `var my_audio =` This will create a variable to hold your audio file. The name of the variable is `my_audio`.

4. From the Effects tab, drag and drop the whenKeyDown **event listener** to your code.

5. Inside the whenKeyDown event listener, type the code: `my_audio.source = "retro.mp3";` Click on the video canvas. The code above will make the song change to `retro.mp3` when you press down on any key!

1

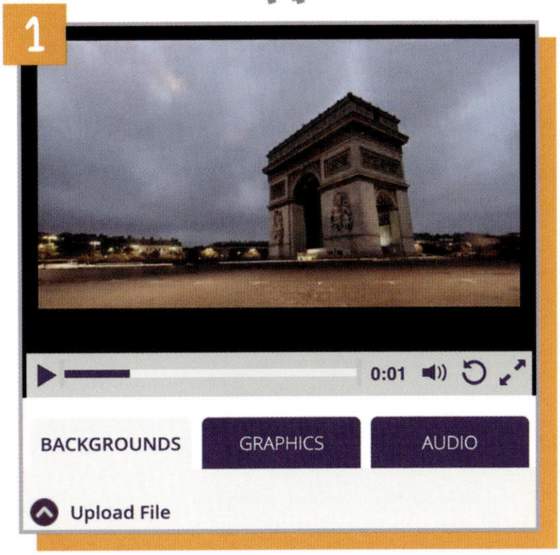

3

```
movie = video();
movie.source = "city_sunset.mp4";

var my_audio = audio("dance.mp3");
```

5

```
movie = video();
movie.source = "city_sunset.mp4";

var my_audio = audio("dance.mp3");

movie.whenKeyDown = function(key) {
  my_audio.source = "retro.mp3";
};
```

6. Above the whenKeyDown event listener block in the code editor, create a new variable called:

```
var my_songs = [];
```

7. Inside the square brackets of the `my_songs` variable, type more audio file names from the Audio tab. Separate the music file names by commas and enclose each file name in quotation marks. You've made an array. An array can hold a list of tunes!

8. Below the array, create a new variable called index:

```
var index = [0];
```

6

```
movie = video();
movie.source = "city_sunset.mp4";

var my_audio = audio("dance.mp3");
var my_songs = [];

movie.whenKeyDown = function(key) {
  my_audio.source = "retro.mp3";
};
```

7

```
movie = video();
movie.source = "city_sunset.mp4";

var my_audio = audio("dance.mp3");
var my_songs = ["dance.mp3", "electronic.mp3",
"retro.mp3", "rock.mp3", "funk.mp3"];

movie.whenKeyDown = function(key) {
 my_audio.source = "retro.mp3";
};
```

8

```
movie = video();
movie.source = "city_sunset.mp4";

var my_audio = audio("dance.mp3");
var my_songs = ["dance.mp3", "electronic.mp3",
"retro.mp3", "rock.mp3", "funk.mp3"];
var index = [0];

movie.whenKeyDown = function(key) {
  my_audio.source = "retro.mp3";
};
```

9. Change the audio source inside the whenKeyDown event listener to the name of your new array of songs, followed by [index]:
`my_audio.source = my_songs[index];`

10. On a new line inside the whenKeyDown event listener, add:
`index++;`
This will make the list of audio files in your array play in order one time.

11. To have the array play more than once, change `[index]` from step 9 to `[index%my_songs.length]`
Now you can DJ the night away with a never-ending playlist!

9

```
movie = video();
movie.source = "city_sunset.mp4";

var my_audio = audio("dance.mp3");
var my_songs = ["dance.mp3", "electronic.mp3",
"retro.mp3", "rock.mp3", "funk.mp3"];
var index = [0];

movie.whenKeyDown = function(key) {
  my_audio.source = my_songs[index];
};
```

11

```
movie = video();
movie.source = "city_sunset.mp4";

var my_audio = audio("dance.mp3");
var my_songs = ["dance.mp3",
 "electronic.mp3", "retro.mp3",
   "rock.mp3", "funk.mp3"];
var index = [0];

movie.whenKeyDown = function(key) {
  my_audio.source =
     my_songs[index%my_songs.length];
   index++;
};
```

HACK IT!

The punctuation // indicates the start of a note. Notes are ignored by JavaScript. So, anything you write after these symbols are just reminders to yourself of important details.

SANDBOX FOR FREE CODING

Test your newfound skills and let your imagination guide your next amazing story!

CODING SKILLS

- imagination
- exploration

PROJECTS:
qrs.lernerbooks.com/Storytelling

1. Click on a video clip from the Backgrounds tab.

2. Mix, match, and experiment with image and video options from the Effects tab.

3. Read through the Reference tab to see the different ways you can create cool coding projects with Vidcode. Cut and paste suggested snippets of code to see how it affects your work.

4. Click the "Save" button at the bottom of the code editor to save your project and come back to it later.

5. Click "Publish" to put your new project online for others to admire!

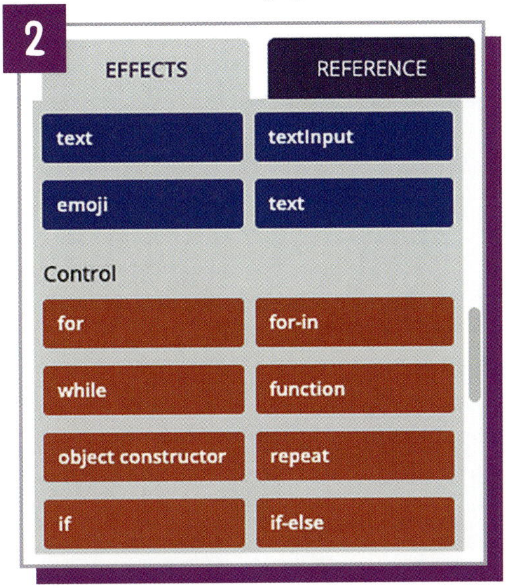

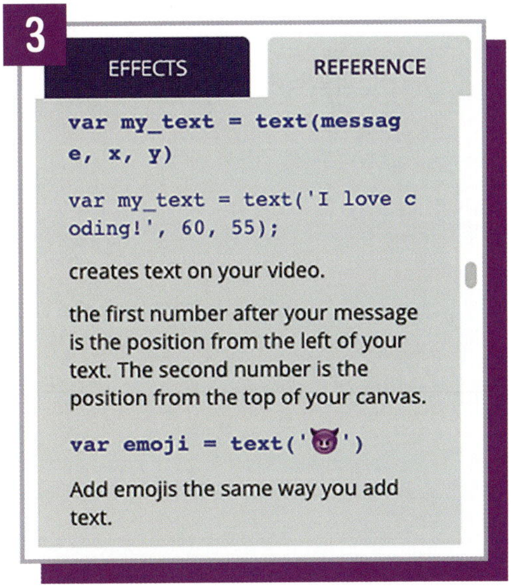

CODING TIP!

Imagination is essential to becoming an expert coder. Have fun coding and creating awesome projects to share with friends and family!

CONCLUSION

You are a coder who can now speak JavaScript! Take your skills to the next level by working on the Vidcode platform to continue learning by doing.

Have fun making awesome coding projects!

MORE ABOUT VIDCODE

Vidcode is a coding platform and curriculum that teaches computer science, object oriented programming, web programming, design, & JavaScript. The magic behind Vidcode has an academic term—computational media! This means learners can upload photos, illustrations, videos and audio—and manipulate them with code right away, allowing tweens and teens to connect computer programming to the media they interact with every day. Vidcode was founded by three women—an engineer, an educator, and an artist. We're now a global team that supports schools, libraries and individual learners in leveraging our 300+ hours of online coding tutorials.

Visit www.vidcode.com for more information.

GLOSSARY

argument: a specific value that acts as input to a function

array: a list of values

canvas: the top-right screen in Vidcode where your coding is displayed as pictures

code editor: the white box where your code is written

coding: writing instructions for a computer. Computer programs are coded.

event listener: a program or function that waits for an event to occur. A mouse click or a key press is an event.

filter: a computer instruction that layers effects over an image or video

frame: an individual image in a sequence of images from a video or animation

function: a set of instructions to do a specific task

loop: in programming, a sequence of instructions that continually repeat

narrative: a series of connected events that make up a story

object: a container or an entity

opacity: the measure of how see-through something is

operator: a symbol that tells a computer to perform a mathematical, logical, or comparison operation

pixelate: to change an image into pixels. A pixel is a square of color on a computer screen. Images can be made up of thousands of pixels.

programming: building or amending a specific set of ordered commands for a computer to perform. A programmer writes computer programs.

programming language: a set of words and rules for giving instructions that a computer understands

property: a value that belongs to an object

sequence: the order in which the commands are executed by a computer

thumbnail: a very small version of an image

value: data such as numbers or text that can be manipulated by a computer program

variable: a container that can store a value

INDEX

PHOTO ACKNOWLEDGMENTS

The images in this book are used with the permission of: © andresr/iStockphoto, p. 30; © Anna_Om/iStockphoto, pp. 4 (boys), 28 (boys); © FatCamera/iStockphoto, pp. 4 (boy), 23; © fizkes/iStockphoto, p. 5 (girl); © Vidcode/Mighty Media, Inc., pp. 4, 5, 6, 7, 8, 9, 10, 11, 12, 13, 14, 15, 16, 17, 18, 19, 20, 21, 22, 24, 25, 26, 27, 28, 29.

Cover Photos: © Henrik_L/iStockphoto; © Vidcode/Mighty Media, Inc.

Design Elements: © Alexey Bezrodny/iStockphoto; © artishokcs/iStockphoto (pixel background); © filborg/iStockphoto (browser window); © Mighty Media, Inc. (blocks)